Skin Tag R

C000193909

Natural Remedies for Removing Skin Tag Safely

Melk Joe

Copyright © 2020 Melk Joe

All rights reserved. No part of this publication may be reproduced, distributed, or transmitted in any form or by any means, including photocopying, recording, or other electronic or mechanical methods, without the prior written permission of the publisher, except in the case of brief quotations embodied in critical reviews and specific other non-commercial uses permitted by copyright law.

ISBN: 978-1-63750-076-7

Table of Contents

Introduction

Have you decided that your skin tag is bothering you; As long as the tag is not infected or near a delicate area, you can remove it yourself at home.

Pores and skin tags are painless, non-cancerous growths on your skin; they're linked to your skin by a little, slim stalk called a *peduncle*. Epidermis/skin tags are regular in men and women, especially after age group 50; they can show up anywhere on the body, though they're common within places where your skin layer folds like the:

- Armpits.
- Groin.
- Thighs.
- Eyelids.
- Neck.
- Area under your breasts.

As pores and skin tags are usually safe, removal is generally for visual or aesthetic reasons. Large epidermis/skin tags, especially in areas where they could rub against something, such as clothing, jewelry or pores, and skin, may be removed credited to irritation.

This book is a best choice for you to learn how to remove skin tag from home.

Chapter 1

What are Pores and Skin Tags?

Epidermis/skin tags are flesh-colored growths that form around the skin's surface; they suspend from a thin little bit of tissues called a stalk. These growths are normal, and about 25% of trusted resources of people have at least one epidermis/skin tag.

You'll usually find epidermis tags in folds from the epidermis/skin in these areas:

- Armpits.

- Neck.

- Under the breasts.

- Round the genitals

Less often, epidermis tags can grow within the eyelids.

Epidermis tags don't cause any medical issues; however, they could be unpleasant if indeed they rub against your clothes.

Skin Tag on Eyelid Removal

You don't need to eliminate a skin tag unless it bothers you. If you want to remove epidermis tags for aesthetic reasons, you have several options.

Home Treatments

Some websites recommend using home remedies like apple cider vinegar to remove epidermis/skin tags. However, before you try removing a skin tag yourself using apple cider vinegar, check with your health-care professional. You don't need to injure your very delicate eye area.

If your skin layer tag includes a fragile base, you might be able to tie it off in the bottom with a little bit of dental hygiene floss or cotton; this will need off its blood flow. Eventually, your skin layer tag will fall off. Ask physician again prior to trying this system. Removing epidermis tags with a good base might trigger significant amounts of loss of blood or contamination. You might leave scar tissue formation on your own eyelid.

Medical Procedure and Treatments

You're safest leaving epidermis tag removal to a health-care professional, which is among the techniques physician will use to get rid of the extra little epidermis/skin from your own eyelid; this treatment will remedy your skin tags you have, yet they won't prevent new epidermis tags from turning up later on.

Cryotherapy

Cryotherapy uses extreme chilly to freeze off epidermis/skin tags; a medical doctor will apply water nitrogen to your skin layer for the cotton swab, or with a couple of tweezers. The liquid may sting or melt away just a little when it continues on your own skin layer; the frozen pores and your skin tag will fall off within ten times.

A blister will form in your community where the drinking water nitrogen was applied, that ought to scab over and fall off within a fortnight to per month.

Surgical removal

Another way to remove skin tags is usually to trim them off. A medical doctor will 1st numb the spot, and then remove your skin tag having a scalpel or special medical scissors.

Electrosurgery

Electrosurgery uses warmth to melt off your skin tag in the bottom. Burning prevents excessive loss of blood when the tag is removed.

Ligation

Within a ligation process, physician ties off underneath of your skin layer tag to remove its blood flow. After weekly or two, your skin tag will perish and fall off.

Causes of Epidermis/Skin Tags on Eyelids?

Skin tags are created from a protein called collagen and

arteries, surrounded with a layer of epidermis; doctors don't understand specifically what can cause them because you'll usually find tags in epidermis/skin folds like your armpits, groin, or eyelids, friction from epidermis massaging against epidermis/skin could be engaged.

Individuals who are overweight or obese are certain to get epidermis tags because they have extra epidermis/skin folds. Hormonal changes during pregnancy may also enhance the probability of epidermis/skin tags forming; there may be a link between insulin resistance, diabetes, and epidermis tags.

People have a tendency to get more epidermis/skin tags because they age; these growths often pop-up in middle generation and beyond. Epidermis/skin tags may run in the family; it's possible that one individual inherits an increased probability of obtaining these epidermis growths.

Preventing Epidermis Tags

It's impossible in order to avoid every skin tag, and you may lessen your possibility of obtaining them by maintaining at a healthy weight. Here are some prevention tips:

- Work together with your physician and a dietitian to plan foods that are reduced saturated surplus fat and calories.

- Exercise at medium or high strength for at least 30 mins every day, five occasions a week.

- Keep all epidermis/skin folds dry in order to avoid friction; pat your skin layer completely dry once you shower. Apply baby powdered to epidermis folds like your underarms that have a tendency to trap moisture.

- Don't wear clothing or jewelry that irritates your skin layer. Choose soft, breathable materials like cotton instead of nylon or spandex.

Risk to Consider

You're more likely to obtain epidermis/skin tags in the event that you:

- Are overweight or obese.

- Are pregnant.

- Possess type 2 diabetes.

- Are within your 40s or older.

- Have additional families with epidermis tags.

Chapter 2

Why Epidermis/Skin Tags Occur

Skin tags are constructed of loose collagen fibres and arteries surrounded by epidermis. Collagen is some sort of protein found during your body. Women and men can possess skin tags; they have a tendency to happen in older people and people who are obese or have type 2 diabetes.

Pregnant women may also be more likely to develop skin tags consequently of adjustments within their hormone levels. A lot of people develop them for no apparent reason. Skin tags have a tendency to grow in your skin layer folds, where in fact the epidermis/skin rubs against itself, such as for example around the neck, armpits or groin. That's the reason they have a tendency to affect obese individuals who've spare folds of pores and tags and epidermis/skin chafing.

When to see Epidermis Tags as a Problem

Epidermis tags are harmless and do not usually stress or discomfort. However, you may consider having epidermis/skin tags removed if they're inside your self-confidence, or if indeed they snag on clothing or jewellery and bleed; you'll usually have to pay out to own this done privately. For the reason that skin tag removal is without a doubt cosmetic surgery, which is usually rarely available through the NHS; cosmetic surgery is usually just within the NHS if the problem has effects on your own physical or mental health. Sometimes, epidermis tags fall off independently if the cells have twisted and died from inadequate blood supply.

Removing Epidermis Tags

Do not try to remove an epidermis/skin tag without talking together with your doctor first. When you yourself have epidermis tag that's resulting in problems, consider producing a meeting having a privately practicing GP to obtain it removed. Skin tags could be burnt or frozen off much like how warts are removed; they could also be surgically removed, sometimes utilizing a local anaesthetic.

Freezing or burning epidermis tags could cause irritation

and short-term epidermis discoloration, as well as your skin tag may not fall off, and extra treatment could be needed.

Surgery gets the advantage of obliterating your skin layer tag, but there are a threat of small bleeding. If your skin layer tag is small using a narrow base, your GP may declare that you try to remove it yourself; for instance, they could suggest tying off underneath of your skin layer tag with dental hygiene floss or cotton to remove its blood flow and produce it fall off (ligation); never make an effort to remove large epidermis/skin tags yourself because they'll bleed severely.

Chapter 3

How are Pores and Skin Tags Removed

Tiny skin tags may rub off independently; however, most epidermis/skin tags stay mounted on your own skin. Generally, epidermis tags don't require treatment, if epidermis/skin tags harm or bother you, you may decide to keep these exact things removed.

A medical doctor may remove your skin layer tags by:

- **Cryotherapy:** Freezing your skin layer tag with liquid nitrogen.

- **Surgery:** Removing your skin layer tag with scissors or a scalpel.

- **Electrosurgery:** Burning off your skin tag with high-frequency electricity.

- **Ligation:** Removing your skin layer tag by tying it off with surgical thread to have ability to remove its blood flow. Tiny skin tags may rub off independently; however, most epidermis/skin tags stay mounted on your own skin. Generally, epidermis tags don't require treatment, if epidermis/skin tags harm or bother you, you may decide to keep these exact things removed.

- **Local anaesthesia:** Having small epidermis tags removed doesn't usually require anaesthesia, a medical doctor might use local anaesthesia when removing large or multiple epidermis tags; you can also try natural treatments to remove skin tags (These include tea tree oil, apple cider vinegar, and lemon juice). Remember that there's no medical evidence to assist these remedies.

It's will's thought to try and remove epidermis/skin tags on your own; many articles offer DIY instructions so you can get rid of epidermis tags by tying them off with string or applying a substance peel from the lime, which isn't very ideal and recommended. Even inside a sterile environment, eliminating epidermis/skin tags might lead to loss of blood, burns, and contamination. It's easier to let a medical doctor handle the task.

How to Identify Epidermis Tag

The principal way to identify a skin tag is from the peduncle; unlike moles and several other epidermis growths, epidermis/skin tags suspend off your skin layer by this small stalk.

Most epidermis tags are small, typically smaller than 2 millimetres in proportions; some can form as large as many centimetres. Epidermis tags are smooth touches. They could be natural and circular, or they could be wrinkly and asymmetrical. Some epidermis/skin tags are threadlike and resemble grains of grain.

Epidermis/skin tags could be flesh-colored; they could also be darker compared to the encompassing skin credited to hyperpigmentation. If an epidermis/skin tag becomes twisted, it might turn black credited inadequate blood flow.

What Causes Epidermis Tags?

It's unclear just what causes epidermis/skin tags. Simply because they usually get to epidermis folds, friction may tend included. Skin tags are constructed of arteries and collagen surrounded by an external layer from the epidermis/skin.

Associated with a 2008 research, the individual papillomavirus (HPV) could be considered one factor in the introduction of epidermis tags. The analysis analyzed 37 epidermis/skin tags from various sites of the body. Results demonstrated HPV DNA in almost 50 percent of your skin layer tags examined.

Insulin resistance, which can result in type 2 diabetes and prediabetes, possibly also are likely mixed up in introduction of epidermis tags. Individuals who have an insulin degree of resistance don't absorb blood sugar effectively from your bloodstream. Regarding a 2010 research, the existence of multiple epidermis/skin tags was connected with insulin degree of resistance, an increased body mass index, and high triglycerides.

Epidermis tags will also be a common side-effect of pregnancy; this is due to carrying a child hormones and gaining weight. In rare cases, multiple epidermis tags can be viewed as a sign from the hormone imbalance or an endocrine problem.

Epidermis/skin tags aren't contagious; there may be a hereditary connection. It isn't uncommon for multiple families to maintain these things.

Risky Facts to Consider

You may be at an increased risk of getting skin tags in the event that you:

- are overweight

- are pregnant

- have got family that has epidermis/skin tags

- have got insulin resistance or type 2 diabetes

- have got HPV.

Epidermis tags don't become epidermis cancer. Irritation may occur if certainly they rub with clothing, jewellery, or another epidermis/skin. Shave with caution around epidermis tags.

Shaving off epidermis tag won't cause long-term damage, though it might stress and prolonged loss of blood.

When to See a Doctor

Since some moles could be cancerous, it's easier to have your skin layer tags analyzed by physician; other skin circumstances such as for example warts and moles can resemble skin tags. Your physician can identify epidermis/skin tags. They'll most likely do this through an obvious exam. If indeed they have any questions about the analysis, they may possibly also execute a biopsy.

Treatment

As epidermis tags are often safe, removal is normally for visual or aesthetic reasons. Large epidermis/skin tags, especially in areas where they could rub against something, such as for example clothing, jewelry or pores, and skin, could be removed credited to irritation.

Removing a large skin tag through the facial skin or beneath the hands could make shaving easier.

Surgery

The next methods can be employed:

- Cauterization: Your skin layer tag is usually burned off using electrolysis.

- Cryosurgery: Your skin layer tag is frozen off

employing a probe containing water nitrogen.

- Ligation: The blood flow to your skin layer tag is interrupted.

- Excision: The tag is slice out having a scalpel.

These procedures should only be performed using a dermatologist, or specialist health-care professional, or a similarly trained medical expert. Epidermis/skin tags around the eyelid, especially those near the eyelid margin, might need to exist removed by an ophthalmologist, or specialist vision doctor.

Removing epidermis tag in the home isn't frequently suggested, because of a risk of loss of blood and possible infection. However, small tags could be removed by tying dental floss or little cotton thread around underneath from the tag to remove blood circulation towards the tag.

Over-the-counter solutions

Over-the-counter (OTC) solutions are available at pharmacies; these freeze your skin layer tag, and it will fall off after 7 to 10 times, that may also be purchased online, though it really is preferred that healthcare advice is gotten before using these treatments. These medications become those utilized for wart removal; no evidence removing skin tags encourages even more of them to develop.

Having small epidermis tags removed doesn't usually require anaesthesia, a medical doctor might use local anaesthesia when getting rid of large or multiple epidermis tags; you can also try natural treatments to get rid of skin tags (These include tea tree oil, apple cider vinegar, and lemon juice). Remember that there's no medical evidence to assist these remedies.

It's will's thought to try and remove epidermis/skin tags on your own; many articles offer DIY instructions so you can get rid of epidermis tags by tying them off with string or applying a substance peel from the lime, which isn't very ideal and recommended. Even within a sterile environment, eliminating epidermis/skin tags might lead to loss of blood, burns, and contamination. It's easier to let a medical doctor handle the task.

Chapter 4

Home Remedy for Skin tag removal

Consult a medical practitioner before trying the following methods.

1. A tag removal device

Home remedies are around for skin tag removal, you can buy online and in lots of stores; however, epidermis/skin tags usually do not require treatment and may fall away independently, but medical removal is available.

People use the unit like a medium to acquire blood to underneath from the tag with just a little band; the medical community recognizes this system as ligation, and with out a supply of blood, the cells will die, as well as the tag will drop away, usually within ten days.

2. String

Some people try to achieve ligation having a little bit of oral floss or string, that could be tricky to accomplish this with no support of these devices or another person since when the blood flow has been removed for at least a few days, the tag should fall away; it could be necessary to shrink the string or floss each day.

Make sure you take the precaution of cleaning your skin layer, string, and hands thoroughly in order to avoid infection.

3. Skin tag removal cream

Packages containing cream and an applicator are available in pharmacies. Instructions for a few packages recommend cleaning your skin layer with alcoholic wipes and processing down the tag before applying the cream, to make sure that it's fully absorbed.

The cream might lead to a mild stinging sensation; however, tags should fall off within 2-3 weeks.

4. Freezing kit

A person might usage something containing liquid nitrogen to freeze off skin tags; the merchandise tend to be accessible in drugstores and pharmacies. Make sure you observe instructions because several applications could be necessary before a tag falls away, but this usually occurs within ten times.

The spray shouldn't touch surrounding epidermis, but a

person might want to apply vaseline to the spot round the tag for protection.

5. Tea tree oil

Tea tree oil could be a oil used to deal with several epidermis circumstances, including epidermis/skin tags. People typically apply several drops of the fundamental oil to a cotton ball, that then they affix to your skin layer tag having a bandage; the cotton ball is normally left on your own skin layer for 10 minutes, three times daily. It could take many times or weeks for the tag to fall off.

This treatment ought to be utilized with caution, as tea tree oil may irritate sensitive skin. Never use tea tree oil for tags located across the eyes.

6. Apple cider vinegar

Little research has been conducted around the potency of apple cider vinegar for epidermis tag removal; people often soak a cotton ball in the vinegar and affix it towards the tag using a bandage for 10 minutes many times each day, before tag falls away.

7. Iodine

Anecdotal evidence demonstrates a person might use liquid iodine to remove skin tags.

First,

- Protect the encompassing skin by using vaseline or coconut oil to the spot.

- Soak a Q-tip in iodine and spread the liquid on the tag.

- Cover the region having a bandage before iodine has dry out.

Continue carrying this treatment out two times per day before tag drops off.

8. Cutting

Physician may recommend trimming away the development using a clean knife or scissors; never attempt this with medium or large epidermis tags, as this might result in blood loss. Tags usually measure from several millimetres to 2 inches wide.

Only look at this method if your skin layer tag includes a fragile base; however, scissors and cutting blades ought to be sterilized before and after make use of. Additionally it is necessary to seek professional advice before attempting this system, and never take off epidermis/skin tags located through the entire eye or genitals.

Home cures aren't ideal for epidermis tags that are:

- Located near to the eyes.

- Located in the genitals.

- Very large or lengthy.

- Causing pain, loss of blood, or itching.

Seek treatment in such instances. Listed here are medical means of skin tag removal:

- Cauterization: This demands burning off your skin layer tag. Most tags will drop away after several treatments.

- Cryotherapy: Physician will apply liquid nitrogen to freeze from your tag. Usually, several treatments will be sufficient.

- Ligation: The tag is linked off with surgical thread, to reduce blood flow.

- Excision: An expert make use of a knife to remove the tag.

Skin tag removal is normally considered beauty, which is unlikely to become contained in health insurance.

When to see a Doctor

If epidermis/skin tags that are large, painful, or located in delicate areas, such a person should see a doctor.

Seek prompt treatment if an epidermis/skin tag:

- Bleeds.

- Itches.

- changes kind or appearance.

Chapter 5

How Toothpaste Remove Skin Tags

Toothpaste had not been developed to deal with epidermis tags; however, this won't imply toothpaste cannot remove your skin layer tags, toothpaste will likely contain hydrogen peroxide to aid in the complete tooth whitening process. Hydrogen peroxide will more than merely pearly white teeth though; besides, it dries out your skin layer.

Drying out your skin layer is just what you should do to get rid of epidermis tags; meaning applying toothpaste which includes hydrogen peroxide to your skin layer tags will eventually produce your skin layer tags dry and fall off. However, it is possible to some things you'll want to know before grabbing the toothpaste from your own medication cupboard and rubbing it on your own skin layer tags.

Can I use any Toothpaste?

The one very first thing you should know may be the sort of toothpaste that you will have to use to remove a skin tag; you are unable to just make use of any toothpaste and expect positive results.

First,

- The toothpaste must contain hydrogen peroxide: Toothpaste which includes hydrogen peroxide is usually often sold as "whitening toothpaste" or something such as for example that. Just search for hydrogen peroxide around the ingredient list.

- Not all toothpaste is made equivalent: Many types of toothpaste are manufactured from gel because making them better to employ. Gel toothpaste will likely taste superior to regular toothpaste because it includes artificial tastes. These artificial tastes may have great flavor and make cleaning your tooth bearable; however, they are able to appeal to ants and may have a lower life expectancy impact on your skin layer. More importantly, they may be doing almost nothing to greatly help remove epidermis/skin tags.

For both listed factors, you need to use a typical white toothpaste (not gel) which includes hydrogen peroxide when putting it to your skin layer tags; this will raise the effectiveness and decrease the prospect of any unwanted side effects, and this sort of toothpaste will cost less than gel-based toothpaste.

Will Toothpaste Harm my Skin?

Toothpaste contains hydrogen peroxide that may dry out your skin layer; meaning utilizing it really helps to treat pores, skin tags, and cause lines and wrinkles. Fortunately, you don't have to utilize the toothpaste to your skin layer to eliminate your skin layer tag; it is possible to apply the toothpaste to just your skin layer tag carefully; this will certainly reduce the unwanted side effects of your skin layer toothpaste on your own skin layer.

It's also sensible to use toothpaste (not gel) which includes hydrogen peroxide no artificial flavors as mentioned earlier.

How to Usage Toothpaste on Epidermis ?

- Put just a little amount of toothpaste on your own finger and rub everything over your skin layer tag.

- Do not get any toothpaste on your own skin layer

because it will dry your skin layer and cause lines and wrinkles.

- Do not get worried in the event that you get slightly toothpaste on your own skin layer; make use of a washcloth with lukewarm water and wipe away the surplus toothpaste.

Many people prefer to put a band-aid on your skin tag after applying the toothpaste; performing this isn't a required step, nonetheless it additionally won't have any adverse influence on your skin layer. Instead, it'll keep the toothpaste setup and may prevent it from being wiped away.

Also, many people choose to use toothpaste with lukewarm drinking water before each would go to sleep. This decision is usually your decision. In the event that you applied the toothpaste and your skin tag, then you definately need not avoid it before sleep.

How Long will it take for my Epidermis Tag to Fall off?

The quantity of time it needs for your skin layer tag to fall off depends upon several factors, including how frequently you apply the toothpaste to your skin layer tag and the length of your skin layer tag. A typical sized skin tag, for instance, should ingest regards to weekly to fall off in the event that you regularly apply toothpaste to it; this includes applying prior to going to rest and when you awaken. It's also sensible to utilize it every few hours as soon as you possess a shower.

The progress will be slow initially, nonetheless it will eventually dry and fall off with the required time, so do not get thrilled when it starts to expire and rip it out. You intend to buy to fall off generally as this will make sure that the complete skin tag is dead and will not reappear. Usually do not choose at your skin layer tag, nor draw it off yourself.

Applying a band-aid to your skin layer tag may or may not help; it could have a tendency to dry the tag a little more, however, the actual fact that toothpaste takes a somewhat long time to remove your skin layer tag may be slightly irritating. People usually want to see instantaneous results; however, because of this treatment for work, you will need patience.

Additionally, immediately removing skin tags often leads to permanent scars; you only need to weigh advantages of promptly eliminating your skin tag and utilizing a permanent (small) scar tissue formation versus eliminating your skin tag over weekly and having no scar tissue formation.

Final Thoughts

Toothpaste for epidermis/skin tag removal is a superb choice for those who that really wants to utilize more natural treatments to get rid of a skin tag. Also, it effectively reduces the opportunity of scarring, and it's a means that doesn't involve having to usage any over-the-counter products you are uncertain of. You only need to give it time and energy to work.

You should be careful to never get an excessive amount of toothpaste on your own skin layer. The hydrogen peroxide will dry your skin layer, that may bring about wrinkles. Also, you need to ensure that the growths are simply epidermis tags. If you are uncertain, you should seek the help of your skin layer doctor to remove some other epidermis/skin circumstances that may be more serious.

Additional natural treatments include apple cider vinegar and the use of tea tree oil with coconut oil if you're considering alternative skin tag removal methods. Your skin layer doctor might help demonstrate through your alternatives and answer any questions it's likely you have.

Chapter 6

4 DIY Skin Tag Methods

So you've decided that your skin layer tag is bothering you; So long as the tag isn't infected or near a delicate region, you can treat it yourself in the home. Below, you will see four (4) of the best skin tag removal methods you could test in the comfort and confines of your house; the 4th technique isn't suggested for home use, though many still make an effort to take a look despite its risks.

- Tie your skin layer tag off

- Usage Wart Remover

- Try Essential Oils

- Scratch the tag off

Let's have a look at every one of these techniques for epidermis tag removal comprehensive.

Technique 1: Tying Your Skin Tag off

Tying epidermis tags off is a preferred solution to eliminate epidermis tags inside a medical office; until lately there weren't many selections accessible to do that promptly and securely in the home. You will see devices that do this for you personally which prevent infection; a popular tool that may do this for you personally may be the Tagband Device; with the product that will can be found in epidermis tag removal package, you can simply remove your skin layer tags in the home. Before this product was obtainable, people used dental floss or sewing thread as an opportunity to skin tag removal method. However, both these procedures do leave the spot susceptible to illness. Plus, because sewing thread is indeed razor-sharp, it could leave the spot more hurt and sore. The discomfort can lead to higher potential for disease.

Methods for getting gone Epidermis/Skin Tags by tying them off

- **Proper Analysis:** Continually be sure you talk with physician or dermatologist to make sure that the development can be an epidermis/skin tag rather than wart or one more thing which may be more serious.

- **Clean the spot:** Use a detergent and normal water

on the spot; then pat it dry out and dab on massaging alcohol consumption to sterilize it.

Discover the Stalk: Make sure you are apparent on what area for connecting off.

- **Wrap Dental hygiene Floss Around underneath from the Stalk:** Connect it securely such that it stops blood flow; however, not limited it slashes into your skin layer.

- **Sterilize the spot Daily:** Keep carefully the area carefully clean and protect it having a bandage; check the spot day-by-day and apply alcohol consumption or antibiotic cream to keep it from getting contaminated.

- **Wait with regards to a Week:** Your skin layer tag should fall off during this time period.

Advantages of Tying an Epidermis Tag Off

- You can get it done at home.

- You merely need teeth floss, rubbing alcohol, and

bandages.

- It's not difficult.

- It should be almost painless.

- It's easy and simple skin tag removal method.

Disadvantages of Tying Epidermis/skin Tag

- It might lead to irritation.

- You shouldn't take action on sensitive areas or near your groin.

- You will get an infection.

Technique 2: Make use of Wart Remover

Some wart removers also concentrate on epidermis tags; those that do, often will see through them with only one use. Though not really a clinically proven method on epidermis/skin tag removal, most wart removers contain salicylic acidity or other elements. Anecdotal evidence works together with salicylic acidity as impressive epidermis tag remover.

Additional wart removers contain drinking water

nitrogen; this is actually the same material that doctors use to freeze off epidermis/skin tags; if you work with something similar to nitrogen in it, the tag will eventually change colors and fall off. Some wart removers market themselves as epidermis tag removers; those generally include a whole kit, filled up with cream and applicator.

Methods for removing Skin Tags Using Wart Remover

- **Proper Diagnosis:** Be confident the spot you are coping with is epidermis tags.

- **Clean the spot:** Use a detergent and normal water then pat dry out.

- **Apply Cream:** Make sure to follow instructions around the bundle. Most recommend getting the cream on for approximately 20 minutes.

- **Wash and Pat Dry:** Rinse the spot with normal water to thoroughly get rid of the cream. Pat dry out softly.

- **Keep carefully the Area Clean carefully and**

Guarded: Just a little scab will form post-treatment. Keep carefully the crust carefully clean and guarded until it falls off around three weeks later.

Benefits of Wart Remover to remove Skin Tags

- It's pretty inexpensive.

- It's easy and quick.

- One treatment may take care of them.

Disadvantages of Wart Remover to get rid of Skin Tags

- It could sting if you are using the task.

- Redness may occur afterward.

- All wart removers are developed differently, so you might not discover one which works for you personally immediately.

Technique 3: Try Essential Oils for Skin Tag Removal

Some essential oils are advantageous in treating skin issues, including skin tags; they may be popular due to its antiseptic and antibacterial properties, tea tree oil is particularly helpful for epidermis/skin issues including epidermis tags. Tea tree oil hails from the Southeast Australian coastline; the fundamental oil has many medical health insurance and beauty benefits, including epidermis/skin benefits, coping with fungal attacks, and clearing coughs and congestion.

Tea tree oil is available in lots of big package stores, medication stores, and health food stores. You can also buy it online. When buying tea tree oil, choose a natural and 100% real oil, primarily serving like a skin tag removal cream, tea tree oil works well and natural; however, it should take some time to remove your skin layer tags with tea tree oil. The amount of time is dependent upon the pace of recurrence of using tea tree oil on epidermis tags and the length of your skin tag.

For many individuals, it should take about 2 weeks before noticing a noticeable difference.

Removing Epidermis Tags with Tea Tree Oil

There are numerous answers to try whenever using tea tree oil to get rid of skin tags; those hateful pounds involve combining tea tree oil with additional natural powerhouses like apple cider vinegar.

- Thoroughly clean your skin layer tag and the region surrounding it.

- Soak a cotton ball and water, squeezing to wring out excess water. Add three drops of tea tree oil.

- Apply within the affected epidermis/skin by massaging the cotton softly over the spot for 5 minutes.

- Wash gently with water and pat dry out.

- Continue doing this technique three times every day before tag falls off.

How to get past Skin Tags using Tea Tree Oil and Apple Cider Vinegar

- In just a little bowl mix collectively four drops of excellent apple cider vinegar, five drops of fresh lemon juice, and three drops of tea tree oil. Blend until thoroughly mixed.

- Drop a cotton ball with this blend then wear it your skin tag.

- Maintain the cotton ball towards the epidermis tag for 3 to 5 minutes.

- Wash and pat dry.

- Continue carrying this out twice per daytime for about ten occasions or prior to the tag falls off.

How to get past Skin Tags with Tea Tree oil and Essential Oils

- Combine three drops of tea tree oil with any 1 tsp of any carrier oil like coconut oil, coconut oil, or jojoba oil. In the event that you combine it with coconut oil, it'll are more of an epidermis/skin tag removal cream due to the coconut oil's regularity.

- Drop a cotton ball in to the mixture and connect to the affected area by massaging it gently for a few momemts.

- Wash and pat dry.

- Continue doing this technique prior to the skin tag falls off.

Advantages of using Tea Tree Oil to eliminate Epidermis/skin Tags

Tea tree oil is normally safe to use; even pregnant women and the elderly could use it, as long as physician certifies it first. It's an all-natural skin tag removal solution; the task is usually practically pain-free.

Disadvantages of using Tea Tree Oil to Remove Epidermis Tags

It takes a lot longer to work than various other strategies, although it's somewhat frustrating because you must carry the fundamental oil onto the spot for minutes during each program, and you must apply the required oil often per day.

Technique 4: Scrape Your Skin Tag Off

This technique could be the most effective solution to eliminate skin tags; nevertheless, you ought not do that technique in the home. Trimming or scratching your skin layer tag off yourself is dangerous, and it leaves you susceptible to infections, and if done improperly could cause extreme bleeding. If you are thinking about using epidermis tag take off, see a doctor to do the duty safely.

When to See a doctor for Epidermis/skin Tag Removal

Though many of the home remedies work wonders to remove skin tags, sometimes you should seek a doctor's take care of safe skin tag removal. You should visit your skin layer doctor for epidermis tag if:

- **Your skin layer tag is new:** You ought to have

physician check any more development on your own skin layer before coping with it in the home. In the event that you don't, you may be inappropriately dealing with something a lot more severe like skin malignancy in the home without knowing it.

- **Your skin layer tag is sore or red:** These may indicate your skin layer tag is contaminated. Picking at an epidermis/skin tag helps it be susceptible to bacteria and bacterias that may cause infections. If your skin layer tag shows indicators of contamination, including pain, bloating, redness, or whether it's warm or hot touch, see a doctor at the initial opportunity.

- **At home, epidermis tag removal strategies didn't work:** Some epidermis/skin tags could be especially stubborn to get rid of at home. When you yourself have attempted at-home removal methods, as well as your skin tag remains, see a doctor to help take it off for you.

- **Your skin layer tag is at a sensitive area:** Epidermis tags can form in very delicate areas like near to the groin and on the eyelids. When you yourself have epidermis tag in either of the places, don't wreak havoc onto it yourself; those areas are susceptible and need a doctor's attention for removal.

- **Your skin layer tag is significant**: Removing a significant skin tag might lead to excess bleeding.

- **You have specific medical ailments:** A lot of people with loss of blood disorders and other medical illnesses shouldn't try to the methods for getting rid of epidermis tags at-home methods; performing this might result in a severe lack of blood. When you yourself have a loss of blood disorder or are taking an anticoagulant, see a doctor to treat your skin layer tags.

What to Expect every time a Doctor Removes an Epidermis Tag

If you're in none from the situations mentioned previously, you'll need a doctor to assist you with epidermis/skin tag removal, you may be slightly nervous about the duty. Knowing what things to expect about how exactly to get rid of epidermis tags at a doctor's visit can simplify your worries and relax your nerves.

- **Removing epidermis tags can be executed quickly within a dermatologist's office:** Generally, the appointment begins with your skin doctor inspecting your skin layer tag to be certain that it is only a harmless skin tag. Likely, your skin doctor can be ruling out any indications of contamination aswell; after that the dermatologist will likely clean the spot and use among the following procedures to remove your skin layer tags:

- **Cutting your skin layer tag off:** Most doctors select to eliminate an epidermis/skin tag by slicing it off quickly at work. The task begins with cleaning the spot with an antiseptic solution; regarding the size and location of the epidermis/skin tag, the physician may rub a numbing solution on your own skin layer. Then, the physician runs on the very sharpened tool to slice the tag away. The spot will exist rewashed, and a bandage will be used. You will likely feel almost no pain through the procedure. Most only feel a pinprick feeling; tag removal may necessitate the physician to stitch the producing wound.

- **Freezing your skin layer tag:** Sometimes, a health-care professional will choose to get rid of a skin tag through freezing it off with very chilly liquid nitrogen. In this system, your skin doctor cleans the spot 1st and applies numbing cream. In that case your dermatologist will swab or aerosol just a little amount of water nitrogen in your community. The spot may tingle or melt away slightly. Your skin layer tag should fall off in 10 to 14 days.

- **Burning your skin layer tag:** If reducing or freezing the tag off isn't an option, your skin doctor may burn your skin layer tag off. Your skin layer tag as well as the encircling field will become washed; the physician then runs on the little bit of cable that's warmed using a power current to melt away the stalk from the tag. Heat will help prevent the epidermis tag from loss of blood; the tag should fall off carrying out a procedure.

Generally, epidermis/skin tag removal causes almost no bleeding. However, in the event that you undertake bleed undertaking epidermis tag removal treatment, the physician may apply the cream in order to avoid the loss of blood before the area.

Remember, though epidermis tags could be annoying and unsightly, nearly all period they aren't considered a medical concern, just an aesthetic one. With that said if you choose to have physician, remove your skin layer tag, check with your insurance provider initial. Many won't cover cosmetic epidermis/skin tag removal.

Lightning Source UK Ltd.
Milton Keynes UK
UKHW020629220921
391008UK00012B/806